GERMAN SELF-PROPELLED ARTILLERY IN WC

WESPE

105MM Guns, Alkett Weapons Carrier & Captured Vehicles

Joachim Engelmann

SCHIFFER MILITARY HISTORY

West Chester, PA

Photo Credits:
Bundesarchiv, Koblenz (BA)
Nowarra Archives
Scheibert Archives
Podzun Archives
Munin Publishing Co. (M)

Translated from the German by Edward Force.

Printed in the United States of America.
ISBN: 0-88740-407-3

This title was originally published under the title,
Wespe - Heuschrecke,
by Podzun-Pallas Verlag, Friedberg.

We are interested in hearing from authors with book ideas on related topics. We are also looking for good photographs in the military history area. We will copy your photos and credit you should your materials be used in a future Schiffer project.

Published by Schiffer Publishing, Ltd.
1469 Morstein Road
West Chester, Pennsylvania 19380
Please write for a free catalog.
This book may be purchased from the publisher.
Please include $2.95 postage.
Try your bookstore first.

A "Wespe" battery is loaded on railroad cars for transport; in front are the guns, in back the supply-train trucks. The crew checks the blocking and lashing of the guns, which are almost five meters long. Their width of 2.28 meters just fits on the flatcars.(BA)

Development of Self-Propelled Artillery

Military motorization in Europe began in 1899 with the first tests at the German Imperial Maneuvers from August 28 to September 1 in the Nellingen-Münsingen-Merklingen area, in France during a troop maneuver near Verdun and St. Menehould on September 6 to 18, and in Austria with the authorization of funds for motor vehicle production. In France, where the people were excited about "automobilism", officers considered the idea of an "automobile artillery." The idea of self-propelled guns thus came into being even before an armed and armored automobile was tested at the Imperial Maneuvers of 1906 in Austria, and in 1912 the Technical Military Committee examined and finally rejected a design of an armored vehicle with tracks, made by Oberleutnant Günther Byrstyn (Railway Regiment). Prussia set up a motorized battalion, with Saxon and Württemberg companies, only in 1911, while Bavaria had a motorized company since 1890.

Before the first great British tank attack on both sides of Arras on the western front in World War I, a stormy development took place. In the course of it, Austria made its heaviest guns, from 24 cm howitzers to 42 cm mortars, mobile in a company equipped with artillery towing tractors, driven by either gasoline engines or electric generators. Railroad guns were essentially self-propelled too. In the German Reichsheer, every army gained a foot artillery vehicle park as of February 3, 1916, and on April 12 of that year

a foot artillery ammunition vehicle column; in December of that year they were reorganized so that each of the 236 divisions gained a motor vehicle column. After the experience and action of the last war years, the 9th battery of every artillery regiment in the Reichswehr was equipped with 77mm Cannon 14 as "motor vehicle guns." They were intended, of course, for anti-aircraft use, but still they were the essential forerunners of self-propelled guns, though still with truck wheels.

For almost ten years, the Reichswehr carried out long-term experiments and tests of the good and bad qualities of self-propelled guns, including truck-mounted types. On the one hand, tractors were preferred, on the other hand, self-propelled mounts with a combination of wheels for roads and tracks for necessary off-road use were considered. The main factor opposed to the self-propelled gun was its doubled risk of both engine trouble and gun damage. The question of halftrack or full-track vehicles played a role in this.

In 1930 five special developments were being built, but the money for them was lacking, and the basic concept that the artillery should remain basically horsedrawn remained unchanged.

When the Wehrmacht was expanded, only one of the five units of a division was motorized, and that was a heavy company; in addition, so were the batteries of the Panzer

and light divisions and (motorized) infantry divisions, but all with halftrack motor vehicles, which proved themselves very well technically. Armored self-propelled full-track gun mounts did not exist in 1942, though specialists of the 6/AHA and WaPrüf 6/HWA, and others on up to General Guderian, had promoted them again and again for the artillery of the fast-moving and tank units, as well as for antitank use, offering practical suggestions well based on theory. Only as the second Russian campaign began and the enemy began its massive counterstrokes, did Hitler basically advocate, in a speech on January 23, 1942, the development of self-propelled artillery, in order to fill a considerable gap in weaponry with infantry escort, mobile antitank, assault and tank escort guns, which the Red Army already had.

Self-propelled guns — now armored howitzers — have tracked chassis as gun mounts, to which the lower pivot mount, with a circular track or pivot bearing for the upper mount, are attached. Compared to the older types of mounts, they are very heavy and technically very expensive. On the other hand, they have all the advantages of high mobility, long range, good off-road capability and all-round fire. Their recoil when firing is absorbed by a lock in the suspension; the force of the shot is diverted into the ground by folding or hydraulic jacks, and the stability of the chassis is increased. The mobility of the self-propelled mount depends to a great degree on the weight of the vehicle, its top speed, range, climbing and wading ability, ditch-spanning ability, braking power, length, width and height, friction, elevation and tendency to tip, plus its ground pressure.

The armor does not need to equal that of a tank, since the firing positions of the batteries are not at the height of the combat troops. Even thinner steel plates prevent the penetration of infantry bullets and shrapnel. The lack of an armored cupola considerably decreases the added weight of the armor. Thus all self-propelled gun mounts of the Wehrmacht had open hulls that offered only limited protection for men and guns; artillery under full armor simply did not exist! On the other hand, experience taught that the expense of self-propelled guns was only justified by armor. Thus the technical solution remained unsettled and divided, as well as constantly in need of service.

The tactical requirements of the self-propelled guns grew out of the need to follow the tank attack for long distances, to be able to change positions and be ready to fire again immediately without delay, so as to have their full effect in fast-moving and constantly changing combat. The firing height was considerably higher than that of the towed gun, that of the "Heuschrecke" about 1.38 meters, of the "Grille" about 1.15 meters and that of the "Rheinmetall B" some 1.25 meters. The ammunition carriers also had to be fitted with tracks. The suggestion of having removable guns came later but was not realized in practice; instead it should be attempted, if need be, to use the ammunition carrier and thus have two possible chassis for one gun — too ideal for the fading war effort. Of course

The first example of many improvisations on captured vehicles — more at the end of this book — was this combination of a 105mm lFH 16 mounted on a British Mark IV 736 tank. There were very few of them.

the unloaded gun could have offered a much lower target. Guns on self-propelled mounts were absolutely not to be misused as improvised substitutes for lacking tanks or assault guns.

With no better final solution in sight, Hitler called, as early as April 4, 1942, for faster production of interim solutions by using available mounts or armored vehicles, with the goal of series production as of the spring of 1943. Though the Panzer II chassis was selected on May 13, 1942 as the self-propelled mount for the lFH 18 and Pak guns, with a chassis made of Panzer III and IV components for the sFH 18, just ten days later Oberst Feuchtinger provided 160 "Lorraine" tractors as chassis for sixty each of the lFH 18 and Pak 40, plus 40 sFH 13 guns; they could reach 42 kph, had 12-mm front armor and were used until the end of 1943.

In 1942 Krupp, Ardelt of Eberswalde and Rheinmetall-Borsig took up the development of particularly low full-track vehicles as self-propelled gun mounts, Krupp for the lFH, 128mm gun and sFH as the "Grille" (Cricket) series, Rheinmetall as the "Scorpion" series, and Alkett at Borsigwalde the Geschützwagen (Gw) II and III/IV, which were built of Panzer III and IV components to save time; for example, Gw III/IV had the engine, fuel pump and filter, electrical and drive systems of Panzer III, the running gear, radiator, exhaust system, ventilator and air filter of Panzer IV, plus a few special parts. Krupp planned to develop Tiger or Panther chassis into mounts for the 17 cm gun and the 21 cm mortar, but these were not built. The improvisations necessitated the elimination of all-round fire, removable guns, and the established weight limit.

The Gw II for the lFH 18/2 (Sd.Kfz.124), or "Wespe" (Wasp), was produced by FAMO in Breslau and Warsaw, and the Gw II self-propelled ammunition carrier followed, with 158 being built. Alkett produced the self-propelled lFH 18/40/2 on the Gw III/IV as Gerät 804, while Geräte 807 and 812 were planned for the sFH 18/1 and sFH 18/5. The first delivery date was to be May 12, 1943. But in 1942, 1248 self-propelled mounts of all kinds were produced, with 2657 more following in 1943 — plans for 150 a month were exceeded — and 1248 in 1944 (monthly average: 104), and in January 1945 60 were built, for a total of 5213, even though assault guns and pursuit tanks had priority.

On October 2, 1942 the lFH mounted on the Assault Gun III or 40 and a steel model of a self-propelled mount for the 88mm Pak 43/1 L/71 and sFH 18/1, made of the same parts, were reviewed by Hitler and approved by him; he expected the planned production of 100 of each by May 12, 1943. He was especially satisfied with the body of the sFH on Panzer III/IV as an unarmored self-propelled mount, and also regarded the makeshift solution of the lFH, mounted on a Panzer II chassis and lightly armored, carrying a crew of five and produced mainly by FAMO in Warsaw, as good. It was given the name of "Wespe" (Wasp).

By the beginning of 1944, 346 "Wespen" (self-propelled) were already in service. On November 7 of that year Hitler approved the limitation of the lFH on Gw IV (Sd.Kfz. 165/1), just being produced by Krupp as the so-called Sfl IV b, to only eight vehicles, since this version did not satisfy the requirements that had been set in 1942: greater mobility than the tanks, quick readiness to fire, all-round fire, removable gun, crew protection from shrapnel and machine-gun fire. For half a year he also stubbornly advocated self-propelled mounts for the 17 cm gun and 21 cm mortar using Gw IV and weighing 58 tons.

The review of all available self-propelled artillery in Berlin from January 27 to 30, 1943 resulted in a military, technical and constructive comparison: Krupp's "Heuschrecke" (Grasshopper, Gerät 5-1026) with Krupp's own lFH in a closed, rotating turret had little chance of being put into production; Rheinmetall used the available lFH 43 and mounted it so it could rotate behind an armored hull; Skoda presented the simplest design on the chassis of the Pz "T 25." All three, each in its own way, allowed the gun to be removed. It turned out, though, that unity of chassis with the Pak 44, all-round fire and removability of the lFH could not all be achieved at the same time.

On May 4 of that year, Hitler called for both tests and temporary solutions for the projects and a clear systematizing of the unclear type concepts. Although on August 21 the "Hummel" (Bumblebee) chassis, with restored all-round firing, was selected as the "final self-propelled mount for the lFH", Hitler decided on September 11, on the advice of front-line officers, in favor of developing the lFH 18/40 on the Panzer III/IV (Hummel) and 3-ton towing tractor with all-round fire and removability on a cross mount. Being removable was intended to preserve the usefulness of the gun if the vehicle broke down. For that reason, at the end of April 1944 Hitler also insisted on another review of the developmental types made by Krupp and Skoda, which represented the conclusion of the designs that had been developed since 1942 as a result of the basic requirement, even though he rated the urgency of production

Originally planned only as a temporary solution, the "Wespe" remained the most frequently built German self-propelled light howitzer, since no final solution was found.

behind that of assault guns and tanks.

To separate the development and production of self-propelled guns from the urgent production of tanks, the Artillery Board of the Army Weapons Office contracted from 1942 on with firms that were not involved in tank production to build "weapons carriers" on full tracks with 17 kph marching speed, removable guns that could be moved on wheeled mounts, all-round fire, shrapnel protection and a uniform chassis for various weapons with varying uses — an excess of demands that could only lead to complicated and unnecessarily heavy vehicles. On February 4, 1944, WaPrüf 4, in conference with the firms of Krupp, Rheinmetall and Steyr-Daimler-Puch, which used Panzer IV, III/IV and 38(t) chassis parts, simplified its requirements by dropping the uniform chassis, raising the speed to 35 kph, strengthening the shrapnel protection and making the vehicles usable for both antitank and artillery tasks.

The firm of Ardelt in Eberswalde, which had produced the lightest, lowest and least expensive vehicle at intervals since April of 1944, now became involved in the new vehicle series, so that Krupp, at the request of WaPrüf 4, moved its development to there, which also reduced the danger of air attack. Mr. Egen of the Krupp firm was responsible for the development of the "weapon carrier" series, but he had constant trouble with his superior in Essen, Mr. Wölfert.

Early in November of 1944 Minister Speer reported to Hitler at length on the concepts and condition of the "weapons carriers" as well as, on November 28, on the "Wespe" and "Hummel" as self-propelled gun mounts for the armored artillery. Hitler regarded the "Wespe" and "Hummel" as "extraordinarily important retreat solutions" and ordered their immediate high-speed production, while the universal solution of the "weapons carriers" — as OLt Ardelt had suggested — was to be brought to a quick completion of development and put into immediate production.

This was the situation at the end of 1944; the "light weapons carriers" (with four wheels each) for the 88mm Pak 43 and lFH 18/40 and the "heavy weapons carriers" (with six wheels each) for the 128mm K 54 and sFH 18 were available in three prototypes, and series production as of the spring of 1945 was being prepared for. As of the autumn of 1945, 350 vehicles a month were to be delivered, having a fighting weight of 13.5 tons.

All the parts came from the Pz 38(t) program, and the firms that had previously produced the Panzer IV were to be in charge of production.

The gun was installed on a low platform in the rear of the vehicle and could be removed with a crane. Lightening the load, though, was not possible. The firing height was 1.77 meters, the ammunition supply was 96 rounds, kept safely under the turning stage. Loading was simple. The gun had a full all-round traverse. A "uniform weapons carrier" was even planned for the future. As it turned out, though, the total development never took place.

"Weapons carriers" with leFH 105mm still saw service at Eberswalde and north of Berlin in April and May of 1945. After the first temporary solution using tank chassis, development proceeded via the so-called "Geschütz-Wagen" to the "weapons carriers" without arriving at an end result.

In addition there was the "Hummel" — another interim development. It was a heavy field howitzer (150mm caliber) on the Geschützwagen III/IV chassis.

WESPE
Special Vehicle 124

The "Wespe", fully armored, with a removable light field howitzer, was planned only as an interim solution until ready for service. But since things did not turn out as planned, the "Wespe" remained in service from 1942 until the war ended. By the time its production ended in 1944, 682 of them were built.

In 1944 nine light armored howitzer units were in the west before D-Day. The top road speed was 40 kph, still a satisfactory achievement for such an outmoded chassis.(1 x BA)

Haftpfl b 9.9.44
Anstrich [Spies flicker] V.V.W. Gastell 9.9.46
Zum R A W am 9.8.46

24,5

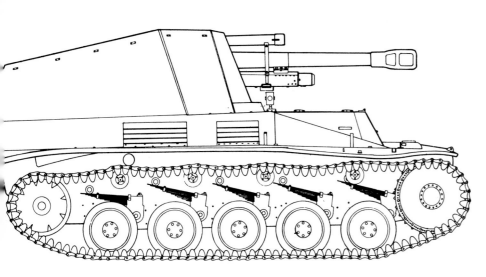

Above: The quarter-elliptic suspension and the fighting compartment angled backward high above the water-cooled six-cylinder engine are typical. At the front, the fighting compartment offers shoulder-high protection; in back, only hip-high.

Right: The barrel of the lFH with gun cradle and pneumatic recuperator ran through a simple slit in the angled front plate but was moved backward so that the driver's compartment could remain unchanged; here the driver is ready to march with the intercom turned on.

Every gun had a camouflage net — generally lashed over the gun barrel. As can be seen by the driver's cap emblem, this "Wespe" belongs to a Waffen-SS battery.(BA)

Left page: Charging the battery is finished, the part of the train that carries supplies is coupled on. The transport leader walks along the train checking everything, and the soldiers look eager and await his comments. The superstructures are covered with canvas.(BA)

This "Wespe" is moving through Beauvais, France, in the direction of Paris in 1944. The crew consists of leader, driver and three gunners. The fighting compartment generally has no weather protection on top — very unpleasant for the men in winter.(BA)

Right: At heights from 2.30 to 3.00 meters, the gun was very nose-heavy in the country, and demanded a lot of ability from the driver and leader. The fuel tank held 200 liters, the engine ran at 2600 rpm. The gun was very hard to camouflage but very mobile and protected from shrapnel.

Below: The 105mm field howitzer made by Rheinmetall-Borsig, its most important parts shown here, had an elevation field of 42 degrees, but at first a traverse of only +/- 20 degrees, so that for major changes the driver had to aim the whole vehicle. All-round fire was made possible later.

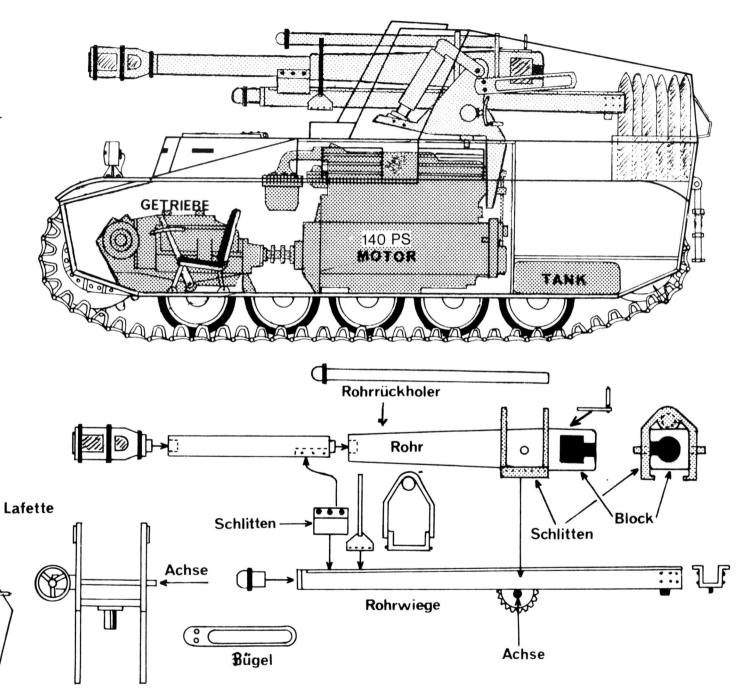

GETRIEBE

140 PS
MOTOR

TANK

Rohrrückholer

Rohr

Block

Schlitten

Schlitten

Lafette

Achse

Schlitten

Rohrwiege

Achse

Bügel

13

Left: A "Wespe", apparently in North Africa (note the soldier's tropic helmet)? It is not known that these guns ever saw action there.

Right page: The data sheet of the "Reich Minister for Armaments and War Production", dating from 1944, gives authentic information on all production figures of Sd.Kfz.124.

Left: The off-road speed decreased to 24 kph. The climbing ability was 30 degrees, climbing ability 42 degrees, ground clearance 34.5 cm, not sufficient for a tank but quite sufficient here.(BA)

Stoffgliederung 21

Geheime Kommandosache!

le FH 18/2 auf Fgst. Pz Kpfw II (Sf)
(Sd. Kfz. 124)

Blatt G' 365

Dringl.-St.: —

Technische Daten:

Gesamtgewicht des Fahrzeuges (Gefechtsgewicht)   11 t

Motor HL 62 TR   140 PS

Höchstgeschwindigkeit   40 km/Std.
Mitgeführte Kraftstoffmenge 200 l (einschl. Reservetank)
Fahrbereich mit einer Kraftstoff-Füllung:
   Straße   220 km; mittl. Gelände   150 km
Grabenüberschreitfähigkeit   1,70 m

Besatzung Fahrer + 1 + 3

Länge   4,81 m, Breite   2,28 m
Höhe mit Aufbau   2,30 m

Bordmunition   32 Schuß f. le. F.H. 18

Bestückung: a) ~~Turmwaffen~~   1 le. F.H. 18 M
   b) ~~Bugwaffen~~
Abfeuerung : Handabfeuerung
Optisches Gerät: a) ~~Turmoptik~~   Rbl. F 36
   b) ~~Kugeloptik~~
   c) Fahreroptik
Funkgerät (normale Ausstattung) Fu. Spr. F., Bordspr. Anl.
Panzerung: Front 18 mm   Seite 15 mm
   Turm 10 mm   Dach 15 mm
Geschütz: Schichau Elbing. Krupp Markstaedt.
   Menck & Hambrock, Hamburg
Kette 108 Glieder, Kettengewicht 385 kg

Rohstoffbedarf	Fe	Mo	Cr	W	Mg	Sn	Cu	Al	Pb	Zn	Ni	Kautschuk (Reifen usw.)
f. 1 Stck. i. kg												

Preis RM	Kfz 43228,— Waffe 16400,—	Durchschn. Fertigungszeit ~12 Monate	Arbeitsstunden

Fertigungsfirmen: Fertigung eingestellt.

The elevation field of the "Wespe" was 42 degrees, the traverse 40. The firing height was 1.94 meters.

15

Ready in position at the edge of a town in the west, November 1944.

The name of "Wespe" was given to the gun unofficially when it was delivered to the troops in 1942; self-propelled artillery was to have insect names. In January 1944 Hitler banned the name, and it remained that of Sd.Kfz.124.(BA)

Above: The bow shows that the turret of the outmoded Panzer II made by Alkett was replaced by a simple box-shaped superstructure, open on top and in the back. The spare track part in front also affords protection. At right, the next gun's battery is being charged.

Upper right: The crew consisted of five men; that of the Panzer II numbered only three. The three gunners were also trained as radiomen and drivers.

Right: With a ground pressure of only 0.82 kg/sq.cm, the gun could wade 92 cm deep and cross ditches 1.70 meters wide. With 14.75 HP/ton, its power-to-weight ratio was rather high.

Until August of 1944, 682 "Wespen" were made by FAMO in Warsaw, specialists in rebuilding.

Above: The "Ammunition Tank II", sister of the "Wespe." carried 90 rounds, as it had additional ammunition space in place of a howitzer. Since the "Wespe" could only carry 32 rounds on account of its large crew, the ammunition carrier was its constant companion, with two of them per battery. It could be armed with a howitzer using its own equipment.

Left: With its gun and machine gun ready, the vehicle moves into the firing position prepared by Richtkreis II.(BA)

Right page: While on its way forward to its firing position, the armored howitzer battery fords a shallow stream.

Above: Driving in the dust of the southern Russian steppes.

Right page: The "Wespe" is ready to fire. At the ordered distance the ordered height is assumed, to be followed by the command to fire.(2 x BA)

Above: A "Wespe" of Panzer Artillery Regiment 76 (6th Panzer Division) before Operation "Citadel."

Upper left: There are pauses between the bursts of fire. Empty shell cases and cartridge boxes lie on the ground, and ammunition is ready near the gun.(BA)

Left: A "Wespe" of the I./SS Panzer Artillery Regiment 5 "Wiking" rolls past a shot-down T-34 tank in the Caucasus in the summer of 1942. Although the gun was just a rebuilt old tank, too heavy for its running gear, it had considerable success and ranked among the best self-propelled artillery guns.

Above: "Wespen" of the I./SS Panzer Artillery Regiment 9 (SS Panzer Division "Hohenstaufen") in bivouac during the summer of 1944. The division was formed in 1943 as an armored grenadier division and reorganized a year later. The photo shows clearly how the camouflage net was used. It would have been better, though, to add some shrubbery.(M)

Right: Though other types were tried, the "Wespe" remained a reliable, tested weapon that gave good service in all theaters of war for more than three years,

The Panzer divisions of the army had at least 76 lFH batteries using "Wespen" in their 38 light armored howitzer units, and a total of, at most, 85 batteries counting armored brigades and special units. Planned as tank escort artillery, the "Wespe" remained the best-known and most important artillery vehicle, although is was developed only during the war and by improvisation.

The Wehrmacht artillery never had genuine "armored howitzers" with full armor protection. Unlike the successful tank production under wartime conditions, here no final technical solution was forthcoming.(BA)

Above: The weight of its shells was 14.81 kg. It could fire six different types of shells.

Upper left: After Rumania fell on August 24, 1944, the 8th Army could stabilize the front at the edge of the Siebenburg Carpathians again. Here light armored howitzer batteries watch over their armored division's retreat routes into the high Carpathians south of Kronstadt, an ideal assignment for these guns.

Left: A "Wespe" battery of one of the eight light armored howitzer units serving in East Prussia in 1945 awaits the command: "Achtung - Fire!"

A self-propelled battery of the armored artillery regiment "Grossdeutschland" moves into position in the area south of Obojan on July 5, 1943, to attack the "Citadel" to the north.

Above: A "Wespe" with damaged tracks, captured by the Americans on the western front in April of 1945.

Upper left: The glow of night firing of a "Wespe" battery shows the typical silhouette of the gun and its crew.(BA)

Left: A "Wespe" battery of the I./Panzer Artillery Regiment 92 of the 20th Panzer Division in the 9th Army's attack to the south during Operation "Citadel" early in July 1943. It is in firing position near Gnilez north of Kursk, where the armored artillery proved itself very well.

After more than 10,000 rounds, the barrel of a "Wespe" was replaced. The barrels with cradle and pneumatic recuperator, i.e. the upper mount, are lifted from an 8-ton towing tractor by an armory troop's crane and inserted precisely, as shown here in Italy in 1944.(BA)

Left: A "Wespe" battery with white winter camouflage, in snow bound firing positions.

Below: An armored artillery regiment (1944) had a diversified profile:

1st Unit: Armored howitzer unit with 2 light armored howitzer batteries, each of 6 self-propelled light Field Howitzer 18/2 (Wespe), and 1 heavy armored howitzer battery of 6 self-propelled heavy field howitzer 18/1 Hummel.

2nd Unit: 1 howitzer unit with 2 light howitzer batteries, each of 6 light Field Howitzer 18 (motorized tractor) or 3 light howitzer units, each of 4 light Field Howitzer 18 (motorized tractor).

3rd Unit: heavy howitzer unit with 2 heavy howitzer batteries, each of 4 heavy Field Howitzer 18 (motorized tractor) and 1 100mm cannon battery of 4 Cannon 18 (motorized tractor).

The regiment had 42 guns. The armored observation battery consisted of one light measuring, one sound measuring, and one weather squadron, plus a printing troop. Each battery had three armored observation cars.

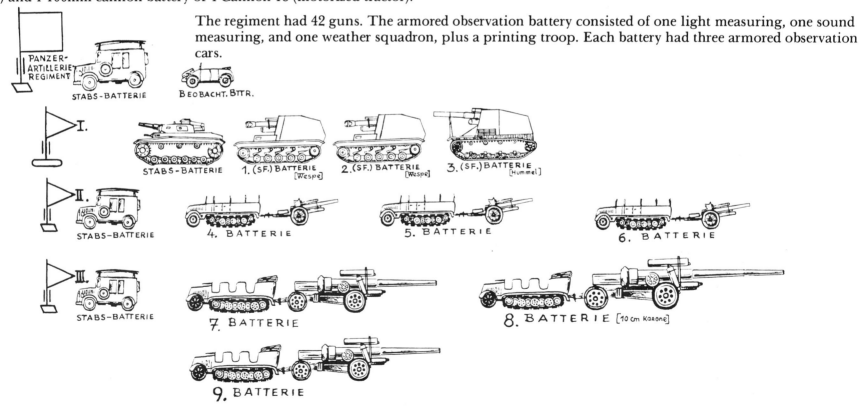

PANZER-ARTILLERIE REGIMENT
STABS-BATTERIE   BEOBACHT. BTTR.

I.
STABS-BATTERIE   1. (SF.) BATTERIE [Wespe]   2. (SF.) BATTERIE [Wespe]   3. (SF.) BATTERIE [Hummel]

II.
STABS-BATTERIE   4. BATTERIE   5. BATTERIE   6. BATTERIE

III.
STABS-BATTERIE   7. BATTERIE   8. BATTERIE [10 cm Kanone]

9. BATTERIE

HEUSCHRECKE

The plan was always to give the guns of the armored artillery full armor protection — including above and in back — and to make them removable. A number of firms were contracted with to produce models. The one made by Krupp on the chassis of the Panzer IV proved to be the most promising. It was given the name of "Heuschrecke" (Grasshopper). But since the chassis was needed even more urgently for assault guns, anti-aircraft guns and such, only a few of them were made.

Thus the "Wespe", originally built as only an interim solution, took on a particular importance for the German armored artillery.

Attempts to mount the 100mm Cannon 18 of the 3rd unit of the armored artillery regiments on self-propelled vehicles failed in their first tests.

Upper right: A Krupp model of a Panzer II chassis lengthened by one road wheel, with a removable light armored howitzer.

Right: Krupp created two different "Heuschrecken." This one has the shortened chassis of Panzer IV (three return rollers instead of four, and six road wheels instead of eight). It was designated IV b.

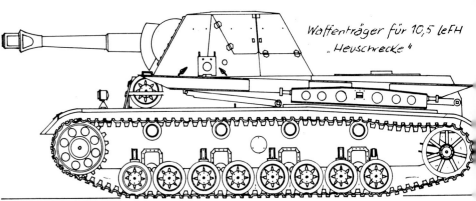

Waffenträger für 10,5 leFH "Heuschrecke"

Above: The other "Heuschrecke" was built on the Geschützwagen III/IV chassis, recognizable by its eight road wheels. As can be seen by the swinging arms, the turret and gun of this model could be removed. This combat vehicle was also called "Heuschrecke 10."

Upper left: Another photo of the same vehicle. This gun could not be removed, but the turret rotated.

Left: This photo shows "Heuschreke 10" with its four return rollers. Only eight examples of the two types were built.

The advantages of this light 105mm field howitzer were its full armor, rotating turret and removable gun.
Thus it could do more than the present-day armored artillery.

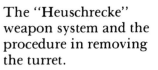

Marching position   Lifting   Swinging

The "Heuschrecke" weapon system and the procedure in removing the turret.

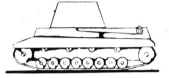

Setting down   Firing   Ammunition carrier

Lower left: A "Heuschrecke" from the rear. The two big wheels on the stern were for the limber (see right page, below).

Turret on limber   Towing   Turret limber from rear

Right page: It can be seen on this page that the removing arms looked somewhat different in the final design from the simple drawing at upper left. The process was certainly not simple and probably not worth the expense. Thus all new armored guns were made without this requirement.

The upper photo shows the removal process with the turret lifted; the turned turret is shown below in marching position on the limber.

Geheime Kommandosache!

le.F.H.18/1 (Sf)/GW IV b
(Sd.Kfz.165/1)

Blatt G 361

Dringl.-St.:

Technische Daten:

Gesamtgewicht des Fahrzeuges (Gefechtsgewicht)   17 t

Motor   Hl 66   PS   188

Höchstgeschwindigkeit   45 km/Std.

Mitgeführte Kraftstoffmenge 410 l (einschl. Reservetank)

Fahrbereich mit einer Kraftstoff-Füllung:

     Straße   250 km; mittl. Gelände   150 km

Grabenüberschreitfähigkeit   2,1 m

Besatzung   Fahrer + 1 + 3

Länge   5900 mm, Breite   2870 mm

Höhe mit Aufbau   2250 mm

Bordmunition   60 Schuss für le FH 18

Bestückung: a) Turmwaffen   le FH 18/1

         b) Bugwaffen   —

Abfeuerung   Handabfeuerung

Optisches Gerät: a) Turmoptik   Rbl. F 36 u. Sf.Zf

         b) ~~Kugeloptik~~

         c) ~~Fahreroptik~~

Funkgerät (normale Ausstattung) Fu. Spr. Ger. F u. Bord-Spr. Ger.

Panzerung: Front 20 mm   Seite 15 mm

         Turm 20/14,5 mm Dach —

Kette 99 Glieder, Kettengewicht 450 kg

		Ni	Kautschuk
			116,3

Rohstoffbedarf	Fe	Mo	Cr	W	Mg	Sn	Cu	Al	Pb	Zn
f. 1 Stck. i. kg	33 t				0,15	1,2	135,1	238	33,3	66,4
o. Waffe, Optik, Funk										

Preis Rm	Kfz 103 468,-	Durchschn. Fertigungszeit	Arbeitsstunden
	Waffe 16400,-	12 Monate	

Fertigungsfirmen:

Above: Along with producing self-propelled mounts for the light field howitzer, there was a parallel development, the 105mm Cannon 18, likewise on the Panzer IV chassis. Krupp also built it.

Left: Changing to the III or IV b chassis meant that the chassis was one and one-half to two times as heavy, longer and more expensive. But the range and amount of ammunition carried were almost doubled.

Right page: There were two examples of this model produced, and they even saw service on the eastern front. There they were nicknamed "Fat Max" by the soldiers.

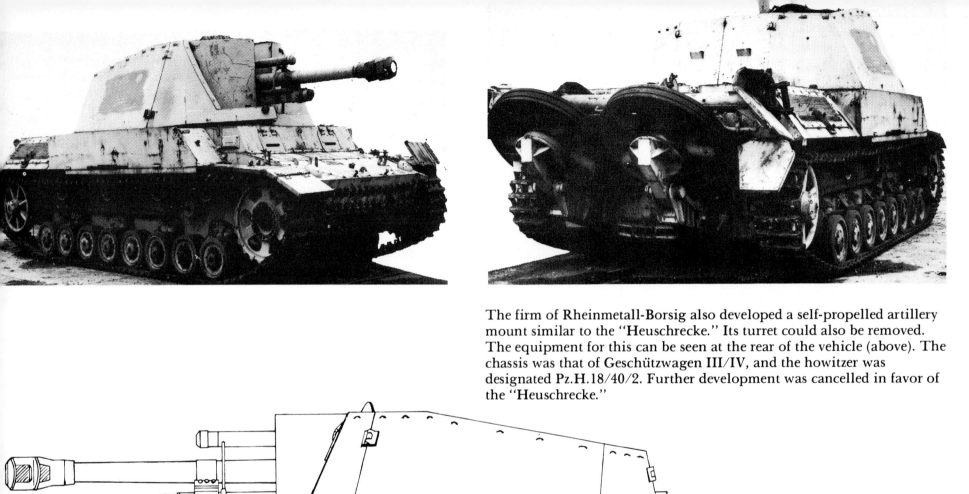

The firm of Rheinmetall-Borsig also developed a self-propelled artillery mount similar to the "Heuschrecke." Its turret could also be removed. The equipment for this can be seen at the rear of the vehicle (above). The chassis was that of Geschützwagen III/IV, and the howitzer was designated Pz.H.18/40/2. Further development was cancelled in favor of the "Heuschrecke."

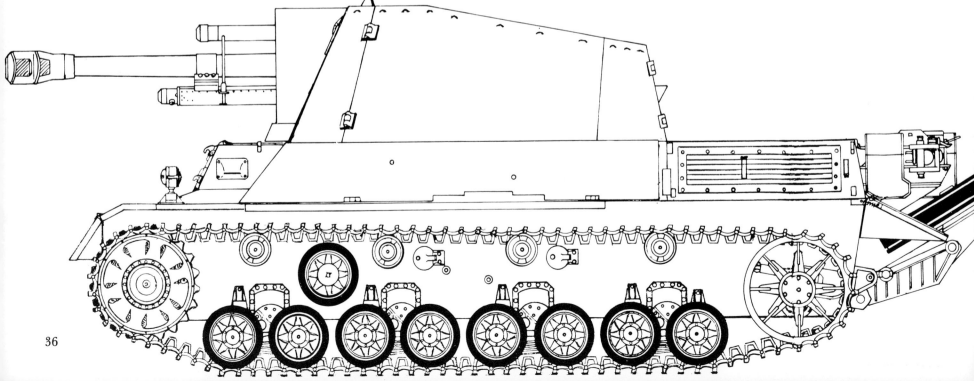

WEAPONS CARRIERS

The many attempts to use available chassis of the Panzer troops as weapons carriers for the artillery and antitank forces always required compromises, had many disadvantages and proved to be too expensive. Thus completely new, simple weapons carriers were designed. They used only some individual parts from already available vehicles, especially the 38(t) tank. Their development was the logical continuation toward a balance of weapon and chassis.

Below: The light uniform weapons carrier made by Rheinmetall-Borsig — shown here with an 88mm Pak 43. Below and at right are the Krupp-Steyr prototype — also with the 88mm Pak 43. Both were also intended to carry the light 105mm field howitzer. Series production did not take place, as the war ended; thus only prototypes existed. The Krupp weapons carrier was also designated Gerät 587 and Geschützwagen 638/26.

Above: A look at the operating compartment of Weapons Carrier II.

Upper and lower left: Above is Weapons Carrier I by Ardelt, with four road wheels. Below is Weapons Carrier II (by Krupp), with six. The first is shown with a 105mm lFH 18/40, Weapons Carrier II with the sFH 18. But the latter was also intended — as seen in the upper picture (view of operating compartment) — to take the lFH 18/40. In this case it was designated as Gerät 578, Geschützwagen 638/21 and medium uniform weapons carrier.

Like the light uniform weapons carrier and Weapons Carrier I, it had a four-man crew.

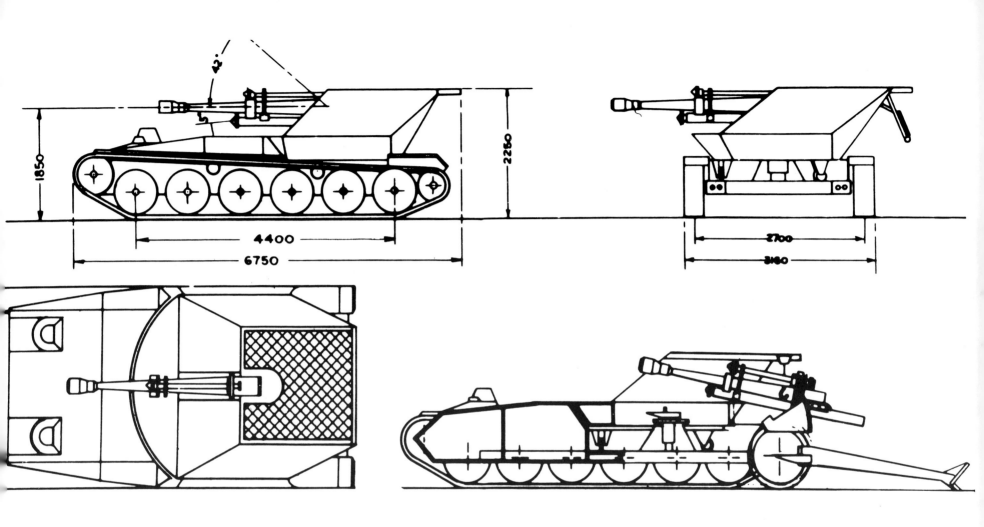

This is how removing the gun of Weapons Carrier I (by Ardelt) was intended. These drawings also show that the superstructure could turn.

Upper left and above:
For the sake of completeness, the installation of a 105mm light gun (L/32) on the Borgward B III (VK 302) special tractor is shown here. The body and gun in the two photos are made of wood. 28 of these combinations were built in 1941 and 1942.

Left:
This Assault Gun III, Type G, is armed with a 105mm assault howitzer 42. It was also called Special Motor Vehicle 142/2. The gun was almost always used in direct fire situations, and larger 75mm weapons.

Mounted on Captured Vehicles

Under the pressure of their ever-increasing need to get armored artillery onto the battlefield, and what with the too-late beginning to develop combinations of their own equipment (Wespe, Heuschrecke) or completely new war vehicles (weapons carriers) — like the Panzerjäger troops — the Germans mounted guns (here the light 105mm field howitzer) on captured vehicles. The vehicles utilized were chiefly French.

In addition to the combination mentioned at the beginning of this volume, with a light field howitzer 16 on a British chassis, there were also the following makeshift combinations, listed here by type and numbers:

— FCM   12
— B2   16
— 39H   48
— Lorraine tractor   60

all of them gave only poor engine performance and were thus, and in other ways, more "desperation moves" than even the "Wespe."

The whole development of German armored 105mm artillery on self-propelled chassis was dominated by the "too late" and "lack of decision" factors. By the end of the war, only the "makeshift Wespe" was seeing actual service, and it accomplished more than had been expected of it. Giving credit to it and its crews was one of the reasons for this book.

There were only twelve examples of the 105mm light field howitzer 42 (based on the lFH 16) on the French FCM gun chassis. They did not see service in the east, and were used only by the Panzerbrigade West in France. The vehicle was very ponderous, complicated to operate, and too heavy for its weak engine. These vehicles were built in 1942.

On the following page is a photo of one of them. Ten more of these vehicles were equipped with the 75mm Pak 40 in 1943.

This combination of a 105mm lFH 18/3 and the French B-2 tank chassis was also not very successful, as it was too high, slow and ponderous. Sixteen of them were built in all and delivered to the occupying troops in France in 1942. It was not used anywhere else.

The crew was composed of four men. With its heavy armor, the vehicle was three times as heavy (about 33 tons) as the "Wespe." 42 shells could be carried.

Technische Daten:

Dringl.-St.:

Gesamtgewicht des Fahrzeuges (Gefechtsgewicht)   32,5 t

Motor   300 PS

Höchstgeschwindigkeit   25 km/Std.

Mitgeführte Kraftstoffmenge 400 l (einschl. Reservetank)

Fahrbereich mit einer Kraftstoff-Füllung:

     Straße   140 km; mittl. Gelände   100 km

Grabenüberschreitfähigkeit   2,75 m

Besatzung   *Fahrer +1+3*

Länge   7620 mm, Breite   2400 mm

Höhe mit Aufbau   3000 mm

Bordmunition ✗

Bestückung: a) Turmwaffen *le FH 18/3*

     b) ~~Bugwaffen~~

Abfeuerung *Handabfeuerung*

Optisches Gerät: a) Turmoptik *Rblf. 36 + Sf Zf.1a*

     b) ~~Kugeloptik~~

     c) Fahreroptik *Periskop*

Funkgerät (normale Ausstattung) *Fu.Ger.5 (S u.E.); Fu.Ger.2 (E). B*

Panzerung: Front *60 mm*   Seite *55 mm*   *Spr.Ger.*

     Turm ✗   Dach ✗

						Ni	Kautschuk			
Kette	Glieder, Kettengewicht				kg					
Rohstoffbedarf	Fe	Mo	Cr	W	Mg	Sn	Cu	Al	Pb	Zn
f. Stck. i. kg										

Preis *RM*	Durchschn. Fertigungszeit Monate	Arbeitsstunden

Fertigungsfirmen:

Pictures of this self-propelled gun are shown on this page too. In all, 48 of them were built. It was better than the two shown previously. It weighed somewhat more than the "Wespe", had — like all of these combinations — a fighting compartment open at the top and a four-man crew. The rebuilt upper mount is typical of it.

This makeshift version also saw service only in France in 1944; the upper photo was taken in the area south of Caen.(1 x BA)

Left page: The two photos show the German 105mm howitzer on the chassis of the French Hotchkiss 39 H tank, above a 105mm lFH 18 and below a 105mm lFH 16. In this photo, Feldmarschall Rommel is inspecting the crew.

| Stoff-gliederung 21 | **Geheime Kommandosache!** GW LrS für le FH 18/4 | Blatt G 363 |

Dringl.-St.: —

Technische Daten:

Gesamtgewicht des Fahrzeuges (Gefechtsgewicht)   7,7 t

Motor De la Haye   70 PS

Höchstgeschwindigkeit   42 km/Std.

Mitgeführte Kraftstoffmenge ÷ l (einschl. Reservetank)

Fahrbereich mit einer Kraftstoff-Füllung:

     Straße   ÷ km; mittl. Gelände   ÷ km

Grabenüberschreitfähigkeit   ÷ m

Besatzung Fahrer +1+3

Länge   4400 mm, Breite   1850 mm

Höhe mit Aufbau   2200 mm

Bordmunition   20 Schuss für le FH 18

Bestückung: a) Turmwaffen   le FH 18/4

     b) ~~Bugwaffen~~

Abfeuerung

Optisches Gerät: a) Turmoptik   Rblf. 36

     b) ~~Kugeloptik~~

     c) ~~Fahreroptik~~

Funkgerät (normale Ausstattung)   10 Watt (S.u.E.)

Panzerung: Front 12 mm   Seite 9 mm

     Turm SmK-sicher Dach ÷

							Ni	Kautschuk	
Kette	Glieder, Kettengewicht					kg			
Rohstoffbedarf f. Stck. i. kg	Fe	Mo	Cr	W	Mg	Sn	Cu	Al	Pb	Zn

Preis .?//	Durchschn. Fertigungszeit Monate	Arbeitsstunden

Fertigungsfirmen:

On this page are two photos and technical data of the Lorraine-Schlepper (f) gun car, here armed with a 105mm lFH 18. Sixty of them were rebuilt for the artillery. They too were used only in France.

Technical Data

	WESPE	HEUSCHRECKE	WAFFENTRÄGER
Type	Las 100/Gw II	BW	I
Manufacturer	Famo/Ursus	Krupp	Ardelt
Years built	1942-1944	1942-1943	1944-1945
Armor cm: hull front	1.8	3.0	2.0
side	1.5	1.6	1.0
rear	1.5	1.6	
Superstructure front	1.0	3.0	
side	1.0	1.6	
rear	1.0	1.6	
Engine type	Maybach HL 62 TRM	Maybach HL 90/100	Praga EPA
Cylinders	1 in-line	V-12 60 degree	6 in-line
Bore x stroke mm	105 x 120	100 x 106	110 x 136
Displacement cc	6191	9990	7750
Compression ratio	6.5 : 1	6.5 : 1	6.2 : 1
Engine speed rpm	2600	4000	2200
Valves	dropped	dropped	dropped
Crankshaft bearings	8 journal	7 journal	7 journal
Carburetors	1 Solex 40 JFF II	2 Solex 40 JFF	1 Solex 48 FNVP
Firing order	1-5-3-6-2-4		1-5-3-6-2-4
Starter	Bosch BNG 2.5/12		Bosch BPD 3/12
Generator	Bosch GTLN 600/12-1500		Bosch GQLN 300/12
Battery V/A	1 12/120		1 12/100
Cooling	water		water
Clutch	dry 2-plate F&S K 230 K		dry 1-plate
Gearbox	ZF SSG 46 Aphon		Praga-Wilson
Speeds fwd/rev	6/1	6/1	5/1
Steering	MAN/Wilson		Praga/Wilson
Turning circle	4.8 meters		
Suspension	leaf springs		leaf springs
Lubrication	high-pressure		high-pressure
Brakes made by	MAN/Famo		Praga
Operation	Mechanical		Mechanical
Road wheel size	500x100-55 mm		775 mm
Track width cm	188		270
Track length	240		310
Track links		108	
Fuel capacity l	200	500	

	WESPE	HEUSCHRECKE	WAFFENTRÄGER
Caliber length	L/18	L/35	L/30
Shell weight kg	14.81	14.81	14.81
Load	6	9	8 cartridges
Muzzle velocity	470 m/sec	645 m/sec	610 m/sec
Shot range meters	10650	14500	13000
Elevation	42 degrees	45 degrees	
Traverse	40 degrees	360 degrees	
Fire height on mount	1.94 meters	2.1 meters	
Fighting weight	11 tons	23 tons	14 tons
Power/weight HP/ton	12.7	17	
Rounds carried	32	60	
Chassis	Panzer II	Panzer IV	1 WaffTr I
Overall length m	4.81	6.0	6.35
Vehicle length m	4.81	6.0	
Vehicle width m	2.28	3.0	3.16
Overall height m	2.3	3.0	2.25
Track width cm	30	40	46
Pressure kg/sq. cm	0.76	0.89	
Ground clearance cm	34	40	45
Fording ability m	0.8	1.4	
Climbing ability m	0.42	0.75	
Spanning ability m	1.7	2.3	
Road/off-road kph	40/20	45/30	20
Range road/off km	140/95	300/150	150
Fuel consumption l road/off-road	90/135	160/300	125
Horsepower	140	400	125

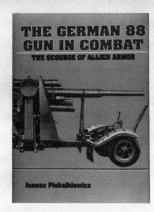
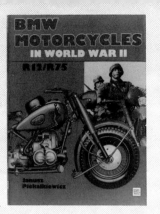

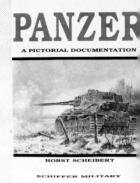